FOREWORD FROM THE CSO

The inaugural Space Capstone Publication, *Spacepo...* is capstone doctrine for the United States Space Force and represents our Service's first articulation of an independent theory of spacepower. This publication answers why spacepower is vital for our Nation, how military spacepower is employed, who military space forces are, and what military space forces value. In short, this capstone document is the foundation of our professional body of knowledge as we forge an independent military Service committed to space operations. Like all doctrine, the SCP remains subject to the policies and strategies that govern its employment. Military spacepower has deterrent and coercive capacities — it provides independent options for National and Joint leadership but achieves its greatest potential when integrated with other forms of military power. As we grow spacepower theory and doctrine, we must do so in a way that fosters greater integration with the Air Force, Army, Navy, Marine Corps, and Coast Guard. It is only by achieving true integration and interdependence that we can hope to unlock spacepower's full potential.

Agility, innovation, and *boldness* have always been the touchstone traits of military space forces. Today, we must harness these traits to pioneer a new Service and a new professional body of knowledge. This capstone doctrine is a point-of-departure toward that goal, not a final adjudication. Given the nascent state of spacepower theory, this publication will inevitably evolve over time as it is applied, evaluated, and refined. Therefore, military space forces are encouraged to read, critique, debate, and improve upon the ideas that follow.

It is an honor to serve with you on this journey.

JOHN W. RAYMOND
General, USSF
Chief of Space Operations

The eyes of the world now look into space, to the moon and to the planets beyond, and we have vowed that we shall not see it governed by a hostile flag of conquest, but by a banner of freedom and peace.[1]

John F. Kennedy, 1962
35th U.S. President

PREFACE
THE SPIRIT OF ORBITAL FLIGHT

Humanity has long been drawn to gaze upward and ponder the mysteries of the heavens. We are defined by our hunger to explore the universe and our unending quest for knowledge of our place in the cosmos. This drive to push back the dark edges of the map and demystify our frontiers is not without conflict. Just as on land, on the seas, and in the sky, the great expanse beyond Earth is now contested.

Once the great powers of the world competed for technological supremacy in outer space to demonstrate the superiority of their societies. To win was to be fastest, highest, farthest, or first. The United States and its Allies firmly won that early space race. Today, however, that competition has evolved, and with heavy consequences. Since this initial competition, the domain of space itself has not changed. The harshness of its environment, its physics, and the vastness of its expanse challenge us today just as it challenged the earliest explorers. Humankind has changed, and our potential adversaries' actions have significantly increased the likelihood of warfare in the space domain. Our destiny as a free country to strive even higher in space remains the same, but the need for security and defense – as only military force can provide – is the stark new reality of our mission.

The benefits to life on Earth from the exploitation of space are inescapable. Never again will our way of living, even at its most fundamental levels, go untouched by the technological marvels orbiting our planet. These systems must be protected from those who would wish to harm them, and in doing so threaten peace and prosperity for free people in every corner of the world. Our draw to explore the unknown, our human love for uncovering our ignorance and replacing it with understanding, will only be possible if we secure the domain of orbital flight.

Our look outward is our look inward; the fundamental desire to know our place among the stars. Human activity and expansion across the domain are not inevitable. The success of these endeavors is only possible if we secure the peaceful use of space, free for any who seek to expand our understanding of the greatest frontier. Today, and far into the future, our military spacepower will be foundational to orbital exploration and development. It is the duty of our Service to fight tirelessly for this mission and take up this awesome mantle the Nation has laid upon us.

MILITARY SPACEPOWER GUIDING PRINCIPLES

- The U.S. desires a peaceful, secure, stable, and accessible space domain. Strength and security in space enables freedom of action in other warfighting domains while contributing to international security and stability. The U.S. must adapt its national security space organizations, doctrine, and capabilities to deter and defeat aggression and protect national interests in space.

- The space domain is the area above the altitude where atmospheric effects on airborne objects becomes negligible. The value of the space domain arises from an ability to conduct activities with unrivaled reach, persistence, endurance, and responsiveness, while affording legal overflight of any location on the earth. Because of these attributes, *spacepower* is inherently global.

- Military space forces are the warfighters who protect, defend, and project spacepower. They provide support, security, stability, and strategic effects by employing spacepower in, from, and to the space domain. This necessitates close collaboration and cooperation with the U.S. Government, Allies, and partners and in accordance with domestic and international law.

⯈ Not only are space operations global, they are also multi-domain. A successful attack against any one segment (or combination of segments), whether terrestrial, link, or space, of the space architecture can neutralize a space capability; therefore, space domain access, maneuver, and exploitation require deliberate and synchronized defensive operations across all three segments.

⯈ As a lean, mission-focused, digital Service, the United States Space Force values organizational agility, innovation, and boldness. Elevating these traits starts with empowering small teams and prizing measured risk-taking as opportunities to rapidly learn and adapt.

Our travels beyond the Earth propel scientific discoveries that improve our lives in countless ways here, right here, at home: powering vast new industry, spurring incredible new technology, and providing the space security we need to protect the American people.[2]

Donald J. Trump, 2018
45th U.S. President

INTRODUCTION

In direct response to the launch of the *Sputnik 1* Soviet spacecraft, pioneers in the U.S. military orchestrated a series of milestones that propelled America's status as the preeminent spacefaring Nation. Launched from Cape Canaveral Air Force Station on 31 January 1958, the U.S. Army spacecraft *Explorer 1* was the first American spacecraft to orbit the Earth.[3] Two months later, the U.S. Navy successfully orbited *Vanguard 1*, demonstrating the first solar powered spacecraft.[4] That same year, the Advanced Research Project Agency, launched, and successfully demonstrated the world's first communications spacecraft. These milestones paved the way for Project Corona, a collaboration between the U.S. Air Force and the Central Intelligence Agency. Launched on 18 August 1960 as part of Project Corona, the *Discoverer 14* spacecraft returned surveillance imagery of over 1,650,000 square miles of Soviet Union territory.[5] Through these successive accomplishments, the U.S. military established that humanity could control and exploit space in pursuit of prosperity and security.

Since these early milestones, the U.S. Army, U.S. Navy, and U.S. Air Force have each developed space capabilities that enhance landpower, seapower, airpower, and cyberpower, respectively. This decentralized context has shaped U.S. military thinking about space for the last 60 years. A product of this institutional structure is that current military theory and doctrine almost exclusively focus on space as an adjunct to other forms of military power without capturing the direct and independent impact space has on U.S. prosperity and security. The doctrine presented in the following chapters elevates *spacepower* as a distinct formulation of military power on par with landpower, seapower, airpower, and cyberpower.

INTRODUCTION

Instead of solely focusing laterally on other domains, an independent conception of spacepower must first recognize the inherent value of the space domain and the tremendous influence space has on U.S. prosperity and security. Chapter 1 of this publication defines the space domain and describes the attributes of orbital flight. The United States harnesses these attributes for exploration, communications, remote sensing, and science. In all cases, the value of orbital flight arises from an ability to conduct activities with unrivaled reach, persistence, endurance, responsiveness, and speed. Chapter 2 applies these attributes by introducing *national spacepower* as the totality of a nation's use of space capabilities in pursuit of national prosperity and security. Under this formulation, space is simultaneously a source and conduit through which a nation can generate and apply diplomatic, informational, military and economic power.

Like any source of national power, the United States must cultivate, develop, and advance spacepower in order to ensure national prosperity and security. *Military spacepower* — introduced in Chapter 3 — exists to preserve that prosperity and security. As a unique form of military power, military spacepower leverages space capabilities to accomplish military objectives in support of national policy and strategy. Military spacepower has deterrent and coercive capacities and the United States Space Force is tasked to provide independent options for national leadership. However, military spacepower achieves its greatest potential when integrated with other forms of military power.

Chapter 4 describes the proper employment of military space forces. Under this framework, military space forces conduct prompt and sustained space operations, accomplishing three Cornerstone Responsibilities — Preserve Freedom of Action in the space domain, Enable Joint Lethality and Effectiveness, and Provide Independent Options to U.S. national leadership capable of achieving national objectives. These three responsibilities form the vital purpose of military spacepower. In order to accomplish these Cornerstone Responsibilities, military space forces must be organized, trained, and equipped to perform *five Core Competencies:* Space Security; Combat Power Projection, Space Mobility and Logistics; Information Mobility; and Space Domain Awareness (SDA). Command and control, and stewardship of the domain, enable military space forces to accomplish these core competencies.

While Chapters 1 through 4 describe why spacepower is vital and how military spacepower is employed, Chapter 5 underscores the men and women who make military spacepower possible. Spacepower requires explorers, diplomats, entrepreneurs, scientists, developers, and warfighters. Military space forces — *protectors of America's space interests* — are first and foremost the warfighters who protect, defend, and project U.S. spacepower. These professionals must simultaneously commit themselves to two demanding professions: warfighting and the mastery of space. This duality blends science and art and forms the core of the purpose, identity, and culture of military space forces.

As a lean, mission-focused, digital Service, the United States Space Force values organizational agility, innovation, and boldness. Elevating these traits starts with empowering small teams and prizing measured risk-taking as opportunities to rapidly learn and adapt. These principles apply equally to operations and day-to-day tasks. The United States Space Force must draw upon these traits to relentlessly advance military spacepower for the Nation.

🜨 🜨 🜨

As the custodian of military spacepower, the United States Space Force has three Cornerstone Responsibilities: Preserve Freedom of Action, Enable Joint Lethality and Effectiveness, and Provide Independent Options. These responsibilities are fed by the five Core Competencies of Space Security, Combat Power Projection, Space Mobility and Logistics, Information Mobility, and Space Domain Awareness. In turn, these Service competencies require specialization in the spacepower disciplines of Orbital Warfare, Space Electromagnetic Warfare, Space Battle Management, Space Access and Sustainment, Military Intelligence, Cyber Operations, and Engineering/Acquisitions.

INTRODUCTION

CORNERSTONE RESPONSIBILITES
1. Preserve Freedom of Action
2. Enable Joint Lethality & Effectiveness
3. Provide Independent Options

Why spacepower is vital to prosperity & security

CORE COMPETENCIES
1. Space Security
2. Combat Power Projection
3. Space Mobility & Logistics
4. Information Mobility
5. Space Domain Awareness

How military spacepower is employed

SPACEPOWER DISCIPLINES
1. Orbital Warfare
2. Space Electromagnetic Warfare
3. Space Battle Management
4. Space Access & Sustainment
5. Military Intelligence
6. Cyber Operations
7. Engineering / Acquisitions

Who is needed, & the skill sets they will employ

All spacepower disciplines are required to execute each Core Competency, just as integration of all Core Competencies is necessary to realize the United States Space Force's Cornerstone Responsibilities. As all these areas are interdependent and interlinked, a space professional can follow the thread from their training in a specific role through to the spacepower disciplines, to the Core Competency it is supporting, to the Cornerstone Responsibility; thus a space professional should see how his/her daily tasking contributes to the safety and security of the Nation.

The Space Capstone Publication is the inaugural doctrine manual for the United States Space Force, providing a basis for training and education, and informs decision-making, mission analysis, objectives, and the development of military space strategy in support of national security, national defense, and national military strategies. In articulating spacepower as a distinct form of military power, this capstone document introduces new terms and concepts to the Department of Defense (DoD), interagency, and Allied communities that represent articulation of an independent theory of spacepower. However, when applicable, this document seeks to maintain alignment with Joint doctrine, allowing the United States Space Force to effectively collaborate with the Joint Force.

The inaugural Space Capstone Publication is the foundational work of the United States Space Force's professional body of knowledge. As the new Service develops and gains operational experience in the emerging strategic environment, so too will its doctrine. Capstone doctrine, in its true form, does not dictate operational responses: it serves as a baseline — founded on historical experience yet propelled by an enduring vision — that allows leaders to tailor their decisions to specific situations. Over time, we can expect additional operational and tactical-level publications to follow as military spacepower is applied, evaluated, and refined.

The proposed doctrine hierarchy for the United States Space Force consists of three levels. There will be a capstone document, under the purview of the Chief of Space Operations (CSO), which articulates the purpose, identity, and values for the Service, DoD, interagency, and Allied communities. The intermediate level of doctrine is operational, providing the organizational support needed for effective military efforts, and further developing doctrine related to the Core Competencies. The final level of doctrine is tactical, which will be referred to as Tactical Standard Operations Procedures (TACSOPs). TACSOPs will codify lessons learned, allowing space force members to apply the various disciplines.

The capstone document typically will be reviewed on a 4-year cycle as the principles should not fundamentally change, although this may occur more often in the early years as the United States Space Force's organization and processes evolve. The field command doctrine center will review the operational doctrine on a two-year cycle. If a need is determined for "flash" changes within doctrine, the United States Space Force's doctrine center commander will have the ability to review/update relevant documents, incorporating changes within time periods as short as three months. Finally, TACSOPs will use an ongoing, online, and collaborative process, approving changes at the lowest appropriate command level, allowing the United States Space Force to rapidly adapt and change with its hallmark agility and boldness.

Space Capstone Publication, *Spacepower (SCP)*
Headquarters United States Space Force
June 2020

Dedicated to past, present, and future spacepower pioneers.

CONTENTS

Foreword from the CSO i
Preface ii
Introduction vii
Chapter 1: The Space Domain 1
    Attributes of Orbital Flight 3
    Space System Architecture 5
    Space Operations 5
        Physical Dimension 5
        Network Dimension 7
        Cognitive Dimension 8
    Challenges of Orbital Flight 8
        Barriers to Access, Movement, and Recovery 8
        Hazards of Orbital Flight 9
Chapter 2: National Spacepower 12
    Politics, Policy, and the Instruments of Power 12
    National Spacepower 13
    Unified Space Action 13
Chapter 3: Military Spacepower 16
    War 16
        War's Enduring Nature 17
        War's Modern Character 17
        Coercion and Victory in War 19
        Space Warfare 21
    Value of Military Spacepower 21
    Full Spectrum Military Spacepower 23
        Key Topology in the Physical Dimension 24
        Protecting Access in the Network Dimension 24
        Affecting the Cognitive Dimension 25
        The Electromagnetic Maneuver Space 25

xiii

Chapter 4: Employment of Space Forces 28
Preserve Freedom of Action 28
Enable Joint Lethality and Effectiveness 31
Provide Independent Options - In, From, and To 31
    Employment Conditions 33
Core Competencies of Military Spacepower in the
United States Space Force 33
    Space Security 35
    Combat Power Projection 36
    Space Mobility and Logistics 37
    Information Mobility 38
    Space Domain Awareness 38
Command and Control of Space Forces 40
Stewards of the Domain 43

Chapter 5: Military Space Forces 46
Developing Spacepower Expertise 46
Spacepower Mentality 48
    Space Warfighters 48
    Mastery of Space 49
    Spacepower Disciplines 50
Art of Space Warfare — Breadth, Depth, and Context 53
Leadership 55
Warfighting Readiness 56
Mission Execution 57
Agility — Innovation — Boldness 58

Acknowledgement 60
References 61

CHAPTER 1 | SPACE DOMAIN

There is something more important than any ultimate weapon. That is the ultimate position — the position of total control over Earth that lies somewhere out in space.[6]

Lyndon B. Johnson, 1958
United States Senator

CHAPTER 1
THE SPACE DOMAIN

Space is the domain of orbital flight. Humanity's ability to achieve and exploit orbital flight — sustained motion beyond the Earth's atmosphere based on gravitational trajectories — ushered in the Space Age. Once orbital flight became a reality, humanity was able to exploit the domain's unique attributes. Orbital flight derives value from the unique characteristics of the space domain's physical environment. Thus, an understanding of spacepower must start with an appreciation of the space domain itself.

ATTRIBUTES OF ORBITAL FLIGHT

The space domain is a unique physical environment. Orbital flight derives its key attributes, and hence value, from these characteristics. The most prominent characteristic is the domain's physical medium. Within terrestrial domains, atmospheric density and pressure resist all forms of motion by generating viscous friction. This force is referred to as *drag* and requires the continuous expenditure of energy to overcome. The space environment is a near — though not perfect — vacuum. This greatly reduces friction, allowing objects to move subject to the pull of gravity. Atmospheric density decreases with altitude, thus objects in orbit closer to the Earth must overcome more drag than those farther out. Furthermore, lacking any significant atmosphere to retain heat, the space domain is subject to extreme temperatures. Finally, undiffused solar energy made up of charged particles and high-energy radiation permeates the domain. The varying properties of these phenomena are collectively referred to as *solar weather*.

Diminished or negligible drag creates conditions that can only be fully exploited in space. Because objects in space encounter negligible drag, spacecraft can maintain extreme velocity without propulsion, enabling them to complete an Earth orbit in as little as 90 minutes depending on orbital altitude. Additionally, signals in the electromagnetic spectrum transmit through the near vacuum of space with little distortion or attenuation, but will decrease in strength over distance due to the inverse square law.

Perspective is another defining attribute of orbital flight. Spacecraft achieve altitudes that provide a perspective beyond the obstruction of Earth and other celestial bodies. The earthward view provides a global perspective and allows an observer to see large swaths of terrestrial surface from a single position, and according to international law, to do so legally, unlike terrestrial observation which do not possess the broad perspective attributes of spacecraft and cannot enter another state's land, sea, or air territory without permission. The outward perspective provides an opportunity to observe celestial objects and orbital activity without the distorting effects of Earth's atmosphere. The benefit of perspective makes orbital flight a valuable and distinct vantage point for observing activity on Earth and across the rest of the space domain.

Orbit or Orbital Flight? Satellite or Spacecraft?

An _orbit_ is any path through space an object follows based on the pull of gravity. For example, the Moon is in orbit around the Earth and the Earth is in orbit around the Sun. While orbits are commonly depicted as circular or elliptical paths, orbits can be repeating or non-repeating. _Orbital flight_ (also referred to as spaceflight) is the act of deliberately manipulating gravitationally curved trajectories in order to transverse beyond Earth's atmosphere and through space. Additionally, orbital flight includes _suborbital trajectories_ that travel into space but deliberately reenter the atmosphere before a complete circumnavigation, _geocentric trajectories_ that remain in space for one or more revolutions around the Earth, and _escape trajectories_ that travel beyond the geocentric regime into the gravitational topology of another celestial body.

While any object in orbit is generically referred to as a _satellite_, the term _spacecraft_ refers to an object which has been engineered to be controlled and deliberately employed in order to perform a useful purpose while traveling in, from, and to the space domain. Small natural objects are referred to as _satellites_ while large natural objects that constitute a significant source of gravity are referred to as _celestial bodies_. Debris refers to any spacecraft or artificial satellite (e.g., a rocket body) in orbit that no longer serves a useful purpose.

Thus, in totality, orbital flight enables a range of activities that are inherently remote, wireless, near instantaneous, and global.

The boundaries of sovereign airspace do not extend into space, and the earliest applications of orbital flight established the international norm of unrestricted overflight for all spacecraft. This makes space a shared environment equally open to all members of the international community. Under this arrangement, the scientific and economic potential of the space domain is boundless. As we look to the near future, orbital flight affords access to an immeasurable supply of economic resources. These resources represent untapped economic opportunities that further elevate the value of the space domain and the imperative for orbital flight.

SPACE SYSTEM ARCHITECTURE

The standard space system architecture plays an important role in space operations. In order to exploit orbital flight, all space systems comprise three distinct segments. The *orbital segment* consists of a spacecraft in orbit beyond Earth's atmosphere. Depending on the application, spacecraft can be remotely piloted, crewed, or autonomous. The *terrestrial segment* encompasses all the equipment within the terrestrial domains required to operate or exploit a spacecraft. This includes control stations, antennas, tracking stations, launch sites, launch platforms, and user equipment. The *link segment* comprises the signals in the electromagnetic spectrum that connect the terrestrial segment and the orbital segment. Uplink signals transmit data from Earth to spacecraft. Downlink signals transmit data from a spacecraft to Earth. Crosslink signals transmit data from one spacecraft to another.

Collectively, this model defines a "system-of-systems" that must operate across the physical, cognitive, and network dimensions. A spacecraft is an engineering marvel; however, it provides little value if it cannot be controlled or exploited. The engineered properties of all three segments play an important role determining the capabilities, limitations, and vulnerabilities of space missions.

SPACE OPERATIONS

Space is a unique physical domain, contiguous only with the air domain, but interconnected with all domains. This shapes how humanity accesses and operates in space. Our ability to conduct operations that access, exploit, and defend space rests on simultaneous action across all domains, but also requires serious consideration of the space domain's unique physical, network, and cognitive dimensions.

PHYSICAL DIMENSION

The *physical dimension* of the space domain encompasses the orbital environment and the spacecraft operating within the domain. This dimension starts in the upper reaches of Earth's atmosphere, intersecting and extending beyond the physical location required for sustained orbital flight. Gravity shapes the invisible terrain of the space domain. The perpetual force of gravity makes constant natural motion the defining characteristic of the space domain's physical dimension. No object in orbit occupies a fixed position. Even a geostationary spacecraft "fixed" over an equatorial longitude travels over 164,000 miles in a 24-hour period.

Orbital Regimes

An orbital regime is a region in space associated with a dominant gravitational system capable of capturing the orbit of other objects. Large celestial bodies generate an invisible topology within their gravitational sphere of influence. This gravitational topology is the fundamental physical demarcation between orbital regimes.

Today, the practical limits of the space domain's physical dimension are segmented into three nested orbital regimes. In the *geocentric regime*, Earth's gravity dominates and objects follow orbital trajectories relative to the Earth. The geocentric regime is nested within the *cislunar regime* – the combined Earth-Moon two body gravitational system. Finally, both regimes are nested in the *solar regime* created by the Sun's massive gravitational field. While these three regimes are the primary regimes of human activity today, the solar system is filled with other orbital regimes, each with a distinct topology.

Once placed in orbit, most spacecraft never return to Earth. Military spacecraft are forward employed for the duration of their lifespan. This makes onboard expendables and consumables (e.g. fuel) and spacecraft reliability the primary determinants of mission duration. Terrestrial perspectives on warfare use attributes like *transit time, range, and endurance* to describe the reach of military operations. The orbital perspective of military power describes the reach of military operations based on access windows, revisit rate, mission lifespan, survivability relative to threat systems, and the tradeoffs between time, position, and total energy.

CHAPTER 1 | SPACE DOMAIN

Space was once a sanctuary from attack, but the emergence, advanced development, and proliferation of a wide range of demonstrated counterspace weapons by potential adversaries has reversed this paradigm. Today, space, like all other domains, is realized to be contested due to the increasing threat to orbiting assets by adversary weapons systems. There is no forward edge of the battle area behind which military spacecraft can reconstitute and recover. Spacecraft remain in orbit through peace and war where they are potentially at risk from adversary counterspace capabilities and the hostile space environment.

NETWORK DIMENSION

The *network dimension* of space operations allows users to command, control, and exploit space capabilities through a physical and logical architecture that collects, transmits, and processes data around the world and across the domain. Because of these dependencies, cyberspace operations within this network dimension are a crucial and inescapable component of military space operations and represent the primary linkage to the other warfighting domains. These dependencies can also create avenues of enemy attack that offer lower costs and higher chance of success than orbital warfare within the space domain only.

Nodes and links are the fundamental components of the network dimension. *Nodes* are elements of the space architecture capable of creating, processing, receiving, or transmitting data. Mission ground systems, control antennas, user equipment, space observation sites, and spacecraft payloads are examples of key space domain nodes. *Links* transport data between nodes. In addition to terrestrial networks, the electromagnetic spectrum (EMS) is a vital link for all space architectures. Orbital spacecraft downlink their data, receive commands, and transmit telemetry through the EMS. Additionally, active and passive EMS applications allow sensors to monitor, detect, track, and characterize resident space objects. Because of the prevalence of remote operations, the EMS is the primary conduit through which the control and exploitation of the space domain is achieved.

COGNITIVE DIMENSION

The space domain's *cognitive dimension* encompasses the perceptions and mental processes of those who transmit, receive, synthesize, analyze, report, decide, and act on information coming from and to the space domain. Every domain of human affairs includes a cognitive component. The prevalence of remote operations amplifies the importance of the space domain's cognitive dimension. While space is a physical location, operations are executed and interpreted through virtual stimuli. Ultimately, space systems are tools that extend the ability of an individual or group to perform tasks in, from, and to the space domain. Space systems are not static systems; they are designed, employed, and exploited by thinking agents. All these associated processes include cognitive components that shape and define human activity in the space domain.

CHALLENGES OF ORBITAL FLIGHT

Overcoming the obstacles of the space domain is the fundamental challenge of orbital flight. Barriers to orbital flight and the hazards of space shape the very nature of space operations.

BARRIERS TO ACCESS, MOVEMENT, AND RECOVERY

Constrained freedom of maneuver is a defining attribute of space operations. Due to orbital mechanics, objects in orbit persist in a state of perpetual motion. Just as impactful is the significant energy required to reach orbital altitudes and to maneuver into a different orbit. Due to extreme velocities, the amount of energy required to reach a different orbit may be significant enough to render the option unfeasible or impractical.

CHAPTER 1 | SPACE DOMAIN

HAZARDS OF ORBITAL FLIGHT

A common misconception is that space exists as an empty vacuum. Such a depiction neglects the dynamic and hostile environment of orbital flight. The environment contains numerous physical hazards and presents a dynamic and hostile operational environment. Earth's atmosphere extends well above the lower threshold for sustained orbital flight, expanding and contracting based on changes in solar activity. In this volume of space, atmospheric drag significantly affects orbital flight. Spacecraft operating beyond the protection of Earth's magnetosphere are not impacted by atmospheric drag but are exposed to solar wind. Originating from the sun, solar wind presents a constant barrage of radiation and charged particles capable of severely damaging a spacecraft's physical and electrical components. While solar wind pervades much of the domain, Earth's magnetosphere traps these charged particles, forming the Van Allen radiation belts. Spacecraft transitioning these regions are further exposed to concentrated levels of charged particles and high-energy radiation.

Space debris poses a further risk to the development of human activity in space. Any artificial space object that no longer serves a useful purpose is space debris and constitutes a collision danger to other objects in orbit. Objects that are not intentionally deorbited will persist until the gradual effects of orbital decay terminate their orbital trajectory. Without active altitude maintenance, the amount of time required for a circular Earth orbit to decay into the atmosphere can range from days (less than 250 miles), years (less than 300 miles), decades (less than 400 miles), or centuries (greater than 400 miles).[7] As the concentration of space debris grows, debris-generating collisions become more prevalent, further jeopardizing orbital flight safety and compromising the utility of the domain.

Additionally, in the contested, congested, and competitive space warfighting domain, our potential adversaries continue to develop, test, and proliferate sophisticated weapons. These weapons have the potential to be the most severe risks to orbital flight. To minimize this risk, it is imperative we execute threat-focused space operations that are fully integrated with timely and relevant intelligence, surveillance, and reconnaissance operations.

⋀ ⋀ ⋀

Since the dawn of the Space Age, attempts to define the space domain have focused on the boundary between air and space. This demarcation has been the source of considerable debate. But just as the ocean's rising and falling tides insufficiently express the depth and complexity of the maritime domain, defining space based on a lower physical boundary neglects the domain's vast expanse and dynamic character. Space is more than an altitude. Space is more than just orbital flight; the concept of space operations must span a physical dimension, network dimension, and a cognitive dimension, among others, in order to completely understand the relationships and interlinkages with the other domains. This functional definition is a necessary component of any attempt to capture the complete utility and potential value of the space domain.

CHAPTER 2 | NATIONAL SPACEPOWER

The beginning stages of man's conquest of space have been focused on technology and have been characterized by national competition. The result has been a tendency to equate achievement in outer space with leadership in science, military capability, industrial technology, and with leadership in general.[8]

National Security Council Report 5814/1
18 August 1958

CHAPTER 2
NATIONAL SPACEPOWER

Access to space is essential to U.S. prosperity and security — it is a national imperative. The many benefits our Nation derives from space include mass communications, financial and economic information networks, public safety, weather monitoring, and military technology. Like any source of national power, the United States must cultivate, develop, and protect these benefits in order to secure continued prosperity.

POLITICS, POLICY, AND THE INSTRUMENTS OF POWER

Spacepower is a source and conduit of national power; thus, an understanding of spacepower must start with an appreciation of international power politics. The term *politics* refers to the deliberate and interactive pursuit of power in the international system while *policy* refers to the political aims and objectives of a nation or non-state actor. Politics is a dynamic social system and can be cooperative or competitive. *Power*, by definition, enables influence and control over events, outcomes, and other actors. States pursue power in order to strengthen their ability to achieve strategic objectives.

States leverage *instruments of national power* in order to exert influence and control in the international system. There are four primary instruments of national power: diplomatic power, information power, military power, and economic power.[9] Collectively, the instruments of power represent the tools states employ to achieve national objectives. The proper conduct and application of spacepower must serve policy aims and seek to strengthen all four instruments of national power.

CHAPTER 2 | NATIONAL SPACEPOWER

NATIONAL SPACEPOWER

National spacepower is the totality of a nation's ability to exploit the space domain in pursuit of prosperity and security. National spacepower is comparatively assessed as the relative strength of a state's ability to leverage the space domain for diplomatic, informational, military, and economic purposes. The space domain is a source of and conduit through which our Nation generates and applies all four instruments of national power. In this regard, space is no different from the land, maritime, air, and cyberspace domains. Space exploration strengthens diplomatic power by conferring national prestige and generating opportunities for peaceful multinational cooperation. U.S. space-based remote sensing and communication is an elemental component of the information power required to employ the other instruments of power. On the modern battlefield, military spacepower has become a prerequisite for global deterrence and power projection. The commercial space industry is a rapidly-growing segment of the U.S. economy with limitless potential. The magnitude of these dependencies makes space a vital and inescapable element of U.S. national power. The application of spacepower must serve policy aims and seek to strengthen all four instruments of national power.

UNIFIED SPACE ACTION

Coherent national spacepower is achieved through *unified space action*. Unified space action synchronizes all components of national spacepower, achieving unity of effort in support of national interests. Barriers to access, maneuver, and recovery make it impractical to completely partition the domain among civil, commercial, national, DoD, intelligence, and military actors — the components of national spacepower must coexist. Unified space action harmonizes these components, guaranteeing they reinforce rather than undermine each other.

Unified space action accomplishes more than de-confliction. The components of national spacepower are mutually reinforcing. For example, military spacepower enables a nation to protect and defend space-based sources of economic power while advances in commercial space technology make military space operations more effective and lethal. At the same time, military and economic power generate a robust backdrop for diplomacy, which leverages space activities to communicate with and influence other actors. Space-based information collection strengthens diplomatic instruments by providing reliable methods to verify international agreements and treaties. Because the components of national spacepower are mutually reinforcing, they must be developed and coordinated for a nation to realize the full strategic benefits of national spacepower.

Military space forces play an important role in achieving unified space action. Military space research, development, and materiel acquisition must be closely coordinated with civil, commercial, and national intelligence space programs.

Military operations in space exist to preserve and advance all equities of national spacepower. Furthermore, the vast size of the military space apparatus necessitates that military space forces play a leading role establishing and reinforcing any standards and norms of behavior in the space domain.

⦙ ⦙ ⦙

Today, the entirety of economic and military space activities is confined to the geocentric regime; however, commercial investments and new technologies have the potential to expand the reach of vital National space interests to the cislunar regime and beyond in the near future. As technology marches forward, U.S. military spacepower must harmonize with the other instruments of power to protect, defend, and maintain the Nation's strategic interests in space.

CHAPTER 3 | MILITARY SPACEPOWER

The purpose of military power is to be prepared, and when called upon by the legitimate governing authority, to maximize violence within the constraints and limitations placed upon it.[10]

Dr. Everett Carl Dolman
***Strategy: Context and Adaption
from Archidamus to Airpower***

CHAPTER 3
MILITARY SPACEPOWER

The value of high ground is one of the oldest and most enduring tenets of warfare. Holding the high ground offers an elevated and unobscured field of view over the battlefield, providing early warning of enemy activity and protecting fielded forces from a surprise attack. Furthermore, forces on elevated terrain hold a distinct energy advantage, increasing the efficiency and longevity of military operations. Finally, control of the high ground can serve as an effective obstacle to an opponent's military, diluting combat power by forcing the enemy to dedicate time and resources away from the main effort in order to dislodge an entrenched force.

The space domain encompasses all of these attributes, making military spacepower a critical manifestation of the high ground in modern warfare. *When employed against adversaries, military spacepower has deterrent and coercive capacities—it provides independent options for National and Joint leadership but achieves its greatest potential when integrated with other forms of military power.*

WAR

Military spacepower is inextricably linked to war. Military space forces must operate in this new warfighting domain to contribute to winning our Nation's wars. Thus, war's enduring nature and modern character shape and define military spacepower.

CHAPTER 3 | MILITARY SPACEPOWER

WAR'S ENDURING NATURE

War is socially sanctioned violence to achieve a political purpose.[11] As Clausewitz said, "War is the continuation of politics by other means." As such, no domain in history in which humans contest policy goals has ever been free from the potential for war. In keeping with international law, the United States acknowledges that the use of space is for peaceful purposes, while preparing for the reality that space must be defended from those who will seek to undermine our goals in space.

As a clash of opposed wills, war manifests as dynamic competition. Belligerents act and react to their opponents, each attempting to thwart the other and gain a position of advantage from which to impose their will. Above all, humans, not weapons, fight wars. This human element injects uncertainty, disorder, surprise, emotion, adaptation, and cunning into the conduct of war.

The term *warfare* describes the methods of waging war.[12] The context of war varies and can range from declared hostilities between sovereign adversaries to limited violence between non-state proxy forces. The competition continuum includes a mixture of cooperation, competition below armed conflict, and armed conflict. In any conflict, political aims, policy restraints, and the law of armed conflict shape the intensity of warfare.

WAR'S MODERN CHARACTER

While war's enduring nature persists, the character of war must constantly evolve. Science, technology, and the endless pursuit of military advantage conspire to shape how wars are fought. For most of human history, direct violence was a universal attribute of war's character. Threatening or applying military force against an adversary required direct contact. Physical danger and immediate personal violence remain central attributes for many modern forms of warfare; however, due to the advance of military technology, these conditions are no longer universal elements of war's modern character.

> ## *Joint Functions*
>
> Joint functions provide a common framework for integrating, synchronizing, and directing operations across multiple domains. Joint Publication 3-0, *Joint Operations,* provides a detailed description of the seven Joint functions and their role in Joint operations. *Command and control* is the exercise of a commanders' authority to direct forces in order to accomplish assigned missions. The *information* function is concerned with the collection, dissemination, management, and application of information in order to drive desired battlefield behaviors, disrupt adversary decision making, and support friendly decision making. *Intelligence* supports a com-mander's predictive understanding of the operational environment. The *fires* function is concerned with the employment of weapons against targets in order to create effects. *Movement and maneuver* seeks to place forces in a position of advantage relative to the ad-versary. *Sustainment* activities provides the logistical support mili-tary forces require to maintain their combat capability throughout deployment, employment, and redeployment. *Protection* preserves the effectiveness and survivability of mission-related military and nonmilitary personnel, equipment, facilities, information, and infrastructure.

The same technology that transformed space into an element of national power has also fundamentally altered war's modern character. The seven Joint functions — command and control, information, intelligence, fires, movement and maneuver, sustainment, and protection — provide a systematic framework for understanding the role space plays in this modern manifestation of warfare. The speed, range, and connectivity of modern weapon systems enable belligerents to wage war on a global scale and across multiple domains. In such a conflict, *command and control, intelligence,* the synchronization of *movement and maneuver,* and force *sustainment* must occur on a global scale and rapidly enough to defeat weapons with extreme speed and range.

Achieving these battlefield conditions creates a dependency on the ability to collect, process, fuse, and disseminate *information* on a global scale. One key distinction of warfare in the Information Age is that many weapon systems rely on external sources of information to function. The prominence of information on the modern battlefield has important implications for force *protection*. In addition to protecting fielded forces, modern warfare demands that the belligerents protect the physical and logical lines of communication that enable the rapid exchange of information. Finally, the range of *fires* within war's modern character now includes lethal and non lethal fires, as well as kinetic and non-kinetic fires. The violent and destructive contributions lethal fires make in warfare endures; however, in the Information Age, maximizing the effectiveness and efficiency of these fires requires an architecture that can strike targets with precision and accuracy. Furthermore, information dependencies create an opportunity for non-lethal and non-kinetic fires from all domains that impair an adversary's access to information and attempt to shatter their decision processes and paralyze their fielded forces.

COERCION AND VICTORY IN WAR

All warfare seeks to coerce an adversary. Formally, *coercion* is the threat or application of force in order to induce an adversary to behave differently than it otherwise would.[13] Coercion in warfare takes many forms and can be divided into deterrence and compellence. Deterrence is the prevention of action by the existence of a credible threat of unacceptable counteraction and/or the belief that the cost of action outweighs the perceived benefit[14] — it essentially seeks to maintain a status quo. Through *extended deterrence* and *assurance*, one actor embraces a partner's interests as their own, thus signaling to any potential adversary or rival a resolve to protect these interests. *Compellence*, within recognized legal limits, attempts to forcibly alter or shape ongoing adversary behaviors and objectives until compliance is reached.

The space domain has global reach and therefore offers a global perspective that is persistent, enduring, and responsive. These characteristics enable forces to deliver effects in the space domain and others from greater ranges, far removed from the U.S., with a minimal operational footprint. This ability is a powerful deterrent to potential adversaries as their awareness of the United States Space Force's global reach influences their behavior and decision calculus on an ever-present basis.

In the competitive context of war, victory results when one side successfully imposes their policy aims on a rival through coercion. All forms of military power — including military spacepower — pursue this objective through the threat or application of force. In so doing, the victor compels a change in behavior an opponent would not otherwise choose to pursue. These outcomes can range from limited concessions to the total surrender or overthrow of an enemy government.

There are two fundamental strategies for the successful employment of military force as a political instrument. *A strategy of incapacitation* seeks to make the enemy helpless to resist by physically destroying adversary military capability.[15] This strategy does not require the total destruction of adversary military forces; rather, incapacitation eliminates military resistance as a viable option for a political rival. In contrast to a strategy of incapacitation, a *strategy of erosion* aims to convince an opponent that adopting the demanded change in political behavior will invoke less pain than continued resistance.[16]

A direct approach is rarely the preferred method of achieving victory. Predictable action is easily thwarted. Instead, success in war often depends on finding an indirect approach that forces an opponent into a disadvantaged or vulnerable position. Once such a position is achieved, we can leverage our strength to strike at and exploit our adversary's weaknesses. This continuous struggle to leverage strength on an opponent's weaknesses makes deception a cornerstone of any strategy in war.

CHAPTER 3 | MILITARY SPACEPOWER

SPACE WARFARE

Adversaries compete in space, from space, and to space in order to achieve political aims and impose their will on opponents. War emerges when the threat or application of military force introduces violence as a mode of interaction between adversaries. Military spacepower can deter behavior and enable access to critical information; however, it can also inflict lethal and non lethal violence against an opponent. As a warfighting force, military space forces must steadfastly prepare to prosecute the appropriate amount of violence against an opponent subject to strategic objectives, legal, and policy restraints.

Just like warfare in any other domain, space warfare is a violent clash of opposing wills.[17] Notably, the adversary in space warfare is never a spacecraft or some other inanimate system. Space warfare targets the mind of an adversary and seeks to neutralize their capability and will to resist. Military space forces compete against thinking actors who threaten our Nation's prosperity, security, or political aims. Thus, military space forces must prepare to outwit, outmaneuver, and dominate thinking, competent, and lethal aggressors who are attempting to thwart U.S. actions.

VALUE OF MILITARY SPACEPOWER

Military spacepower is the ability to accomplish strategic and military objectives through the control and exploitation of the space domain. Military space forces are the practitioners of military spacepower who provide a global perspective to the Joint Force. Security, deterrence, and violent competition are the hallmarks of a warfighting force, and military space forces are no different. They shape the security environment, deter aggression, and apply lethal and nonlethal force in, from, and to space.

However, the attributes of the space domain make military spacepower unique from other forms of military power (just as terrestrial forms of military power are unique from spacepower). Space is the only physical domain capable of achieving a globally persistent and legal overflight military perspective of any location on the earth. Military spacepower achieves this global persistence by combining the high-altitude perspective of space with the enduring longevity of forward employed spacecraft and an international legal regime which recognizes overflight of any point on the Earth by spacecraft. This affords unique opportunities for military power.

Orbital flight extends lines of communication into the most desolate and remote areas of human activity. Military forces at every echelon of war capitalize on this perspective to share information beyond their line-of-sight, synchronizing global power projection across all warfighting domains. The practice of legal, unrestricted overflight allows spacecraft to penetrate the most restricted segments of the battlefield. It is through this denied area access that space-based intelligence, surveillance, and reconnaissance (ISR) provides the foundational intelligence to analyze adversary capabilities, courses of action and intent in order to deliver predictive intelligence for Space Domain Awareness (SDA), and Joint Force Commanders' decision-making processes. When fully integrated, these capabilities along with terrestrial and airborne ISR capabilities will deliver game-changing intelligence to protect and defend the space domain. By controlling this ultimate perspective, military forces can monitor and rapidly respond to any contingency around the world before establishing a large in-theater footprint. Thus, the global, legal, penetrating, and persistent attributes of orbital flight make spacepower assertively responsive to emerging threats around the world.

In addition to the earthward perspective, orbital flight provides an unmatched outward perspective of the space domain. Spacecraft can monitor orbital activity without the obscuring effects of dense atmosphere, terrestrial weather or concealed orbital regimes. Some parts of the space domain can only be observed from an orbital perspective, such as the far side of the moon. Any requirement to observe and monitor obscured portions of the space domain can only be accomplished from an orbital trajectory.

On a fundamental level, operating from the high ground of space reduces an adversary's ability to surprise us. It is difficult for an adversary to take offensive action or hide from a perspective that is simultaneously global, legal, penetrating, and persistent. However, the events of 9/11 remind us of the ever-pressing need for vigilance regardless of a perceived advantage. Considering this, the space perspective can guard against tactical, operational, and strategic surprise.

CHAPTER 3 | MILITARY SPACEPOWER
FULL SPECTRUM MILITARY SPACEPOWER

> *Space Operations Across Three Simultaneous Dimensions*
>
> As an example of full spectrum military spacepower, consider a missile warning spacecraft in geosynchronous Earth orbit. A specifically selected trajectory in the physical dimension of the space domain allows a remote sensor to constantly monitor a portion of the Earth's surface for missile activity. Thus, the payload on this spacecraft serves as a critical information node and is part of its network dimension. When a missile event occurs, the spacecraft will collect data on the event and downlink that data to a mission ground station for processing. Processed information will then be disseminated to end users to inform cognitive decision processes. Furthermore, because of this persistent and overt missile warning, adversaries can credibly expect that a strategic attack will be detected, attributed, and reciprocated, thus strengthening the cognitive aspects of deterrence.
>
> This simple example demonstrates the practical design of military spacepower that deliberately integrates actions across all three dimensions of space operations. A holistic and integrated view of the space domain and space operations must dictate how military services organize, train, equip, and present space forces to be employed by the Joint Force.

A complete understanding of military spacepower must encompass the domain's physical, network, and cognitive dimensions, among others. These dimensions are not sequential segments in a serial architecture. Instead, this model of space operations recognizes the simultaneous and interrelated influence all three dimensions have on military spacepower. These dimensions are tightly coupled, and space operations must deliberately consider and affect all three in order to realize the benefits of spacepower.

KEY TOPOLOGY IN THE PHYSICAL DIMENSION

Identifying, seizing, exploiting, and protecting valuable physical locations is a critical component of military power in all domains. Dividing the physical environment by common or advantageous orbital trajectories is a useful method for compartmentalizing operations in the domain. Thus, a systematic understanding of lines of communication (LOCs) and key orbital trajectories (KOTs) allows military space forces to grapple with the vastness of the space domain when planning, executing, and assessing spacepower operations.

A LOC is any route that connects employed military forces with a base of operations and along which supplies and military forces move. Control of critical LOCs enables the timely repositioning, resupply, and reinforcement of military forces within the space domain. Examples of valuable LOCs include spaceport launch trajectories, spacecraft recovery trajectories, minimum energy transfer paths from one orbit to another, and transfer paths from the geocentric regime to the cislunar regime and beyond.

A key orbital trajectory (KOT) is any orbit from which a spacecraft can support users, collect information, defend other assets, or engage the adversary. These critical segments of the domain represent orbits for mission execution and power projection. KOTs can be defined relative to a celestial body (*inertial KOT*), relative to an advantageous energy state (*energy KOT*), or relative to other trajectories (*orbital KOT*). Some examples of an inertial KOT, the most basic form of "key terrain," are Low Earth Orbit (LEO), Medium-Earth Orbit (MEO), Geosynchronous Earth Orbit (GEO) and sun-synchronous orbits.

PROTECTING ACCESS IN THE NETWORK DIMENSION

The network dimension is not static. Rather, it represents a physical and logical maneuver space. Some examples of tactical maneuvers in this dimension include monitoring and defending software, attacking adversary computer systems, duplicating networks, amplifying signals, shifting frequencies, upgrading encryption, and adjusting data pathways. In a military context, links and nodes that comprise the network dimension of the space domain are potential vulnerabilities and subject to attack. Physical or logical attacks against any segment of the network dimension have the potential to isolate a space system from its end user. Thus, mission assurance requires protecting all segments of the network dimension with a whole of government approach.

AFFECTING THE COGNITIVE DIMENSION

Military spacepower shapes and manipulates how adversaries process information, form perceptions, derive key judgements, and make decisions – all aspects of the cognitive dimension. The space domain's remoteness challenges operators to understand physical operations without direct sensory input. Thus, affecting the cognitive dimension involves preserving the ability to *observe* and *orient* within the environment in order to effectively *decide* and *act* from a remote location. Explicitly acknowledging the cognitive dimension of the space domain emphasizes military spacepower as a coercive force despite the limited number of humans who occupy the domain's physical dimension.

The cognitive dimension includes friendly actors, adversaries, and third parties operating at all levels of warfare. For example, space actions can deter adversary aggression based on the credible perception of an unacceptable counteraction, or assure Allied and commercial partners of a secure space domain.

In space warfare, many of the most critical objectives are achieved in the cognitive dimension. Decision superiority, deterrence, dissuasion, compellence, and assurance manifest here. Neutralizing an adversary spacecraft offers limited military value if such actions fail to influence the perceptions or decisions of the enemy.

THE ELECTROMAGNETIC MANEUVER SPACE

The transfer of electromagnetic energy across a spectrum of different energy states — collectively referred to as the *electromagnetic spectrum* (EMS) — impacts all three dimensions of the space domain, defining the very character of space operations. First, most spacecraft harness naturally occurring electromagnetic energy in order to generate power. Additionally, all spacepower applications rely on remote operations. This makes the EMS an important physical maneuver space within which many important network links reside, connecting spacecraft to terrestrial operators, users, and customers. Wireless connectivity through the EMS enables the rapid dissemination of data that in turn impacts the cognitive processes of those who rely on space-derived information to make decisions, generate economic value, or realize a military advantage. Without these applications of the EMS, modern space operations would not be possible.

While the EMS influences all three dimensions of space operations, it also serves as a potential conduit through which space missions can be disrupted or held at risk. Naturally occurring electromagnetic radiation can pose a tremendous risk to spacecraft electronics. Weaponized directed energy can damage a spacecraft or its payloads. Electromagnetic energy can disrupt or deny EMS links, isolating a spacecraft from operators and users. If left unprotected, false data or information can be injected through the EMS into space networks, allowing adversaries to manipulate the cognitive processes of decision makers and space system operators. These vulnerabilities present a tremendous risk to the viability of military spacepower; therefore, military space forces must prepare to exploit and defend the EMS as a weaponized maneuver space.

▲ ▲ ▲

Military spacepower cannot unilaterally win wars, but like landpower, seapower, airpower, or cyberpower, its success, absence, or failure could prove catastrophically decisive in war. Because military spacepower has the potential to be the difference between victory and defeat, it must be viewed with equal importance as military power in any other domain. This observation is the strategic imperative for creating the United States Space Force as an independent military Service capable of maximizing military spacepower as a distinct and vital formulation of military power.

In the long haul, our safety as a nation may depend upon our achieving "space superiority." Several decades from now, the important battles may not be sea battles or air battles, but space battles, and we should be spending a certain fraction of our national resources to ensure that we do not lag in obtaining space supremacy.[18]

**Major General Bernard Schriever, 1957
Commander, Western Development Division,
Air Research and Development Command**

CHAPTER 4
EMPLOYMENT OF SPACE FORCES

Major General Bernard Schriever delivered his prescient description of space superiority eight months before humanity first launched an artificial spacecraft into orbit. Today, military space forces conduct prompt and sustained space operations, accomplishing three cornerstone responsibilities. First, military space forces Preserve Freedom of Action in the space domain. Second, military space forces strengthen and transform the Lethality and Effectiveness of the Joint Force. Third, military space forces provide U.S. national leadership with Independent Options capable of achieving strategic effects. Taken together, these three Cornerstone Responsibilities define the vital contributions of military spacepower and the core purpose of the United States Space Force.

PRESERVE FREEDOM OF ACTION

The United States' ability to project and employ national power is predicated on access to space. Therefore, unfettered access to and freedom to operate in space is a vital national interest.[19] Assuring freedom of operation in space is a fundamental role of the U.S. military, and specifically the United States Space Force. Preserving Freedom of Action describes a strategic condition where a nation or sovereign actor has the relative level of control or ability required to accomplish all four components — diplomatic, informational, military and economic — of their implicit or explicit space strategy. Preserving Freedom of Action becomes an operational imperative in peace and war, and space security becomes a critical mission across the conflict continuum.

CHAPTER 4 | EMPLOYMENT OF SPACE FORCES

> ***Cornerstone Responsibilities of Military Space Forces***
>
> *Preserve Freedom of Action* – Unfettered access to and freedom to operate in space is a vital national interest; it is the ability to accomplish all four components of national power – diplomatic, information, military, and economic – of a nation's implicit or explicit space strategy. Military space forces fundamentally exist to protect, defend, and preserve this freedom of action.
>
> *Enable Joint Lethality and Effectiveness* – Space capabilities strengthen operations in the other domains of warfare and reinforce every Joint function – the US does not project or employ power without space. At the same time, military space forces must rely on military operations in the other domains to protect and defend space freedom of action. Military space forces operate as part of the closely integrated Joint Force across the entire conflict continuum in support of the full range of military operations.
>
> *Provide Independent Options* – A central tenet of military spacepower is the ability to independently achieve strategic effects. In this capacity, military spacepower is more than an adjunct to landpower, seapower, airpower, and cyberpower. Across the conflict continuum, military spacepower provides national leadership with independent military options that advance the nation's prosperity and security. Military space forces achieve national objectives by projecting power in, from, to space.

Freedom of Action requirements are assessed relative to national strategy. At the political level, military freedom of action may be of limited value should the United States lose the ability to employ the other three instruments of power through the space domain. *This insight demands that the military concept of preserving freedom of action in space also supports the other components of national spacepower. Because of this holistic approach, military space forces must maintain a strategic perspective and appreciate domain requirements that extend beyond the use of military force.*

The terms space parity, space superiority, and space supremacy describe relative degrees of advantage between two or more adversaries. *Space parity* describes any condition where no force derives a relative advantage over another at a given time. *Space superiority* is a relative degree of control in space of one force over another that would permit the conduct of its operations without prohibitive interference from the adversary while simultaneously denying their opponent freedom of action in the domain at a given time. *Space supremacy* implies that one side could conduct operations with relative impunity while denying space domain freedom of action to an adversary. Space supremacy is not always desirable, or attainable against a peer adversary, and should not be the unconditional goal of military spacepower. A rival may wish for their adversary to maintain some space capabilities to reduce the probability of a strategic miscalculation or larger escalation in the conflict and to allow certain elements of national infrastructure to continue without disruption, such as medical services and communication between friendly and adversary decision-makers to broker a de-escalation.

The above conditions can be temporary or permanent. Deliberate action can create windows of space superiority or supremacy in support of specific objectives for a specified period of time. Assessing relative control is predicated on SDA. Military space forces must understand the entwined web of U.S. space dependencies in order to determine how to effectively protect and preserve space domain freedom of action. Assessment involves determining how unsatisfied space dependencies impact implicit or explicit strategy. Thus, military space forces must understand both friendly and adversary space dependencies to make operational assessments.

CHAPTER 4 | EMPLOYMENT OF SPACE FORCES

ENABLE JOINT LETHALITY AND EFFECTIVENESS

The ability to control and exploit the space domain is an essential component of modern warfare. Military spacepower allows for the rapid dissemination of information on a global scale. Information can be collected and delivered to austere environments without terrestrial infrastructure. Capabilities such as precision attack, maneuver warfare, strategic warning, and global power projection are fundamental to modern warfare. These capabilities must be protected, and military space forces must rely on military operations in the other domains to do so.

Given the vital and interdependent nature of military spacepower within the Joint Force, military space forces must comprehensively and effectively integrate space capabilities into Joint training, planning, and operations. Maximizing Joint lethality and effectiveness requires a cadre of military space forces that are deliberately prepared to integrate spacepower across the range of national and Joint operations. Said differently, the United States Space Force must be Joint-smart from its inception and it must help produce a space-smart Joint Force.

PROVIDE INDEPENDENT OPTIONS - IN, FROM, AND TO

Because nations can generate and apply national power from space, actions in the domain can directly affect a nation's decision calculus. Therefore, a central tenet of military spacepower is the ability to independently achieve strategic effects. In this capacity, military spacepower is more than an adjunct to landpower, seapower, airpower, and cyberpower. Across the conflict continuum, military spacepower provides national leadership with independent military options that advance the nation's prosperity and security. Military space forces achieve national and military objectives by operating *in, from and to* the space domain.

Operations in the space domain provide a diverse set of options for national leadership. Actions as benign as the launch or repositioning of a space capability can assure international partners or signal U.S. resolve to strategic competitors. Any nation or non-state actor who depends on space as a source of diplomatic, informational, military, or economic power is vulnerable to military spacepower's coercive potential.

The coercive value of military spacepower is not limited to great power competition. Military spacepower may still have a coercive impact on nations or actors who lack significant space dependencies by projecting power from the space domain. The ability to legally transcend the most remote and protected national boundaries provides a unique opportunity to enable lethal and non-lethal effects against terrestrial targets. Additionally, the specter of global vigilance incentivizes adversaries to adapt their behavior and expend time, energy, and resources to mask or obscure sensitive events. By strengthening our ability to detect and attribute malign action, or using space-based National Technical Means for arms control verification, military spacepower makes deterrence more credible. Military spacepower's ability to enhance the lethality and effectiveness of the Joint Force provides an asymmetric advantage that contributes to conventional deterrence.

Finally, military space forces can shape the security environment through space domain collaboration and security cooperation. Building a partner's spacepower capacity provides non-violent opportunities for strategic effects. These relationships promote U.S. interests by encouraging our partners to act in support of U.S. objectives. By strengthening our alliances and military partnerships, security cooperation makes deterrence more credible and effective. Through these actions, military space forces can reinforce the confidence our partners place in their relationship with the United States.

CHAPTER 4 | EMPLOYMENT OF SPACE FORCES

EMPLOYMENT CONDITIONS

Military spacepower is employed across a range of operational conditions. From enduring requirements to emerging challenges, from permissive environments to hostile situations, and from overt operations to covert or clandestine activities, the employment of military space forces must remain responsive to the needs of the Nation. Given the diversity of these conditions, military spacepower requires a broad portfolio of capabilities and systems. Enduring operational requirements demand exquisite capabilities optimized for performance and longevity. Special employment activities may require purpose-built equipment capable of unique modes of employment. In some circumstances, the only way to respond to an emerging requirement on operationally relevant timelines may be through a rapidly-fielded prototype, an experimental capability, or a repurposed research and development (R&D) platform. Through the combined employment of exquisite, purpose-built, and repurposed capabilities, military space forces capitalize on new technology to provide battlespace effects across a range of operational conditions.

CORE COMPETENCIES OF MILITARY SPACEPOWER IN THE UNITED STATES SPACE FORCE

The United States Space Force executes five Core Competencies: Space Security; Combat Power Projection; Space Mobility and Logistics; Information Mobility; and Space Domain Awareness. These Core Competencies represent the broad portfolio of capabilities military space forces need to provide successfully or efficiently to the Nation. *These Core Competencies are taken in aggregate and applied to achieve the three cornerstone responsibilities. The United States Space Force is organized, trained, and equipped to perform these Core Competencies on behalf of the Joint Force.*[20]

Military Spacepower Core Competencies

Space Security establishes and promotes stable conditions for the safe and secure access to space activities for civil, commercial, intelligence community, and multinational partners.

Combat Power Projection integrates defensive and offensive operations to maintain a desired level of freedom of action relative to an adversary. Combat Power Projection in concert with other competencies enhances freedom of action by deterring aggression or compelling an adversary to change behavior.

Space Mobility and Logistics (SML) enables movement and support of military equipment and personnel in the space domain, from the space domain back to Earth, and to the space domain.

Information Mobility provides timely, rapid and reliable collection and transportation of data across the range of military operations in support of tactical, operational, and strategic decision making.

Space Domain Awareness (SDA) encompasses the effective identification, characterization and understanding of any factor associated with the space domain that could affect space operations and thereby impact the security, safety, economy, or environment of our Nation.

CHAPTER 4 | EMPLOYMENT OF SPACE FORCES

SPACE SECURITY

U.S. prosperity and economic security increasingly rely on the peaceful use of space. *Space Security* protects these interests by establishing conditions for the safe and secure access to space for civil, commercial, Intelligence Community (IC), and multinational partners. Space Security is a presence mission that helps assure partners that the U.S. military is positioned to monitor and protect their interests. Ultimately, Space Security seeks to encourage partners, not compel an adversary; however, if necessary, Space Security includes protecting these mission partners from dangerous or illicit actions. In this regard, combat forces provide a deterrent role for Space Security. Space Security may also include sharing information and domain awareness, developing self-protection capabilities, coordinating anomaly resolution support, maneuver de-confliction, EMS monitoring, launch vehicle ridesharing, protecting lines of communication and national space commerce, and building partner capacity through combined training and exercises.

Cooperation and coordination are the defining attributes of Space Security. In order to effectively perform Space Security, military space forces must develop strong connections with their civil, commercial, IC, and multinational partners. Space Security is mutually beneficial. Military space forces ensure a safe and secure environment for our partners. In return, our partners bolster U.S. space capacity while sharing vital information that increases space domain awareness. These strong connections allow all parties to assist each other during crises.

As the range of civil, commercial, national intelligence, and multinational space applications expands in scope and extends farther from Earth, military space forces must prepare to extend Space Security in support of these new U.S. interests.

COMBAT POWER PROJECTION

Combat Power Projection ensures freedom of action in space for the United States and its Allies and, when necessary to defend against threats, denies an adversary freedom of action in space. Combat power – the force available and, when necessary, employed to protect, defend, or defeat threats – is the purest form of military power and serves a distinct role in the pursuit of strategic objectives. Military space forces project combat power for defensive and offensive purposes. Defensive operations enhance control by protecting and preserving U.S. freedom of action in the space domain. When warranted, offensive operations are designed to achieve a relative advantage by negating an adversary's ability to access, or exploit the space domain and are therefore essential to achieving space superiority.

Defensive operations protect and preserve friendly space capabilities before, during, or after an attack. Defensive operations are further divided into active and passive actions. Active defensive operations encompass actions to destroy, nullify, or reduce the effectiveness of threats holding friendly space capabilities at risk. Although this may entail reactive operations after an adversary has initiated an attack, active defense also includes proactive efforts to seize the initiative once an attack is imminent. While active defense directly attempts to interdict the employment of capability against friendly space missions, passive defense attempts to improve survivability through system and architectural attributes. Passive defense measures include spacecraft maneuverability; self-protection; disaggregation; orbit diversification; large-scale proliferation; communication, transmission, and emissions security; camouflage, concealment, and deception; and system hardening across all three segments of the space architecture.[21]

Offensive operations target an adversary's space and counterspace capabilities, reducing the effectiveness and lethality of adversary forces across all domains.[22] Offensive operations seek to gain the initiative and may neutralize adversary space missions before they can be employed against friendly forces. Offensive operations are not limited to adversary counterspace systems and can also target the full spectrum of an adversary's ability to exploit the space domain, which includes targets in the terrestrial and cyber domains.

CHAPTER 4 | EMPLOYMENT OF SPACE FORCES

The space system architecture – space segment, terrestrial segment, and link segment – makes defensive and offensive operations inherently multi-domain. In order to preserve a space capability, all three segments must be protected. Conversely, a successful attack against any segment of the space architecture can neutralize a space capability. Thus, defensive and offensive operations must be employed through an effects-based approach across all three segments. The focus is on a specific outcome or impact to adversary operations.

SPACE MOBILITY AND LOGISTICS

Space Mobility and Logistics (SML) is the movement and support of military equipment and personnel into the space domain, from the space domain back to Earth, and through the space domain. The ability to control and exploit the space domain always begins with physical access to orbit. SML starts with the ability to launch military equipment into the proper orbit in a safe, secure, and reliable manner. During conflict, space launch must be dynamic and responsive, providing the ability to augment or reconstitute capability gaps from multiple locations. Today, SML is largely uncontested, though the history of warfare highlights that this condition will not last. Military forces must therefore prepare to defend physical access to the space domain – a key focus of defensive operations and the need for military space forces to be prepared to project combat power.

Orbital sustainment and recovery is another important application of SML. Already demonstrated in the commercial sector, orbital sustainment will allow military space forces to replenish consumables and expendables on spacecraft that cannot be recovered back to Earth. Orbital sustainment will also enable spacecraft inspection, anomaly resolution, hardware maintenance, and technology upgrades. Orbital recovery allows for the recovery of personnel or military equipment from the space domain. This includes objects such as reusable spacecraft or launch boosters

INFORMATION MOBILITY

Information Mobility is the timely, rapid, and reliable collection and transportation of data across the range of military operations in support of tactical, operational, and strategic decision making. Information Mobility includes point-to-point communications; broadcast communications; long-haul communication links; protected strategic communications; machine-to-machine interfaces; position, navigation and timing; nuclear detonation detection; missile warning; and intelligence, surveillance and reconnaissance.

Global power projection introduces information collection and transportation requirements which space is uniquely postured to deliver. Information Mobility extends lines of communication into the most desolate and remote areas of human activity. Military users capitalize on this perspective to share information beyond their line-of-sight, synchronizing power projection on a global scale and across all warfighting domains.

Information Mobility is a deliberate mission that must be planned, integrated, and tailored with other warfighting requirements. Information Mobility is also a contested capability that must be protected through active and passive measures.

SPACE DOMAIN AWARENESS

Space Domain Awareness (SDA) encompasses the effective identification, characterization, and understanding of any factor associated with the space domain that could affect space operations and thereby impacting the security, safety, economy, or environment of our Nation. SDA leverages the unique subset of intelligence, surveillance, reconnaissance, environmental monitoring, and data sharing arrangements that provide operators and decision makers with a timely depiction of all factors and actors — including friendly, adversary, and third party — impacting domain operations. Furthermore, SDA must be predictive, synthesizing facts and evidence into an assessment of possible and probable future outcomes.

CHAPTER 4 | EMPLOYMENT OF SPACE FORCES

Fundamentally, SDA is a big data challenge. The United States Space Force must have the ability to collect, synthesize, fuse, and make sense of extremely large volumes of data from all sources to ensure the United States Space Force's ability to have domain awareness. As a digital Service, the United States Space Force must leverage its personnel, Allies, civil and industry partners, and big data toolsets to reveal patterns, trends, and associations, especially relating to human behavior and interactions that provide the required SDA.

SDA extends across the physical, network, and cognitive dimensions of space operations. Operating in the physical environment of space requires a timely awareness of space weather, lighting conditions, and gravitational topology. In addition to these natural phenomenon, military space forces must also maintain awareness of spacecraft orbiting in the domain. This includes active spacecraft and debris. Moreover, when tracking active spacecraft, SDA captures more than orbital trajectory. Complete SDA also includes mission related details such as missions, intentions, system capabilities, patterns-of-life, and the status of consumables and expendables.

Awareness of the network dimension must encompass the links and nodes that enable orbital flight and the movement of information in, from, and to the domain. This includes the frequency, location, access, and power of EMS links along with the physical and logical pathways required to transmit information across space architectures. SDA provides insight into key redundancies and chokepoints in the network dimension.

Awareness in the space domain's cognitive dimension encompasses the actors who operate or rely on space systems, along with their decision-making processes, biases, cultural values, and psychological tendencies. Importantly, military space forces must also maintain an awareness of their own decision processes and any associated personal or institutional biases. SDA of the cognitive dimension allows commanders to detect deceit, determine adversary intentions, and act within an adversary's decision cycle.

The practical reality of SDA is that we will not have all possible information all the time. Instead, SDA must be deliberately planned and maintained to ensure the right information is delivered to the right decision maker at the right time. Thus, SDA can be viewed as a self-reinforcing process: SDA helps predict future outcomes and conditions, which in turn drives future requirements for domain awareness.

In addition to the items listed above, there are two overarching areas that enable the effectiveness of these five Core Competencies: Command and Control, and Stewards of the Domain.

COMMAND AND CONTROL OF SPACE FORCES

Traditional models of warfare segment military objectives into three interrelated levels: strategic, operational, and tactical. This model captures the relationship between national objectives and tactical action — tactical objectives are nested within operational objectives, which are nested within strategic objectives.[23]

Command and Control (C2) is the exercise of authority and direction by a properly designated commander in the accomplishment of a mission.[24] Effective C2 ensures unity of command within assigned and attached forces and enables unity of effort between those forces and external organizations. It relies upon the efficient arrangement of personnel, equipment, communications, facilities, and procedures. Ultimately, effective C2 hinges on the communication of intent, delegation of decision-making authority to appropriate echelons, and timely judgments that are informed by higher-level guidance, available battlespace awareness, and operational experience.

The C2 of space operations depends on clearly defined authorities, roles, and relationships. Unambiguous delineation of the chain of command, support relationships between organizations, and levels of delegated control for assigned forces are prerequisites to decentralized execution of space operations. A well-designed command and control scheme specifies the authorities delegated to commanders and functional leaders, whilst still permitting flexible control frameworks. Depending on the operational situation, such control architectures could be organized by orbital regime, along space core competencies, or through the integration of those competencies into composite force packages. A practiced battle rhythm and rehearsed coordination mechanisms between organizations will help ensure the selected command

CHAPTER 4 | EMPLOYMENT OF SPACE FORCES

and control structure fosters unity of effort across disparate military organizations and with other stakeholders, especially the national Intelligence Community.

The nature of orbital flight can induce strategic compression, blurring the distinctions between the tactical, operational, and strategic levels of war. A single spacecraft may support multiple theaters, be difficult to replace, and represent immense national strategic value. This often amplifies tactical action, rapidly propagating the results of a confined engagement into operational and strategic level effects.

As with any operating domain or area of responsibility, the command and control of space forces reflects the distinctive character of space operations and the unique attributes of the space domain's physical dimension. Space C2 requires closing a complex decision cycle, often on compressed timelines, at great physical standoff, synchronization across disparate coordinating organizations, and with efficient management of a limited pool of high-demand, low-density resources. These challenging features entail special considerations for the effective C2 of space forces.

Space C2 is oriented toward tactical action and designed to outpace the adversary by boldly seizing and maintaining the initiative. Overcoming the physical remoteness of space operations requires timely and anticipatory SDA that can only be achieved through a "team-of-teams"[25] approach to C2. Military space forces cannot operate as discrete, isolated units because no single unit possesses a complete operational picture, even for a limited tactical space engagement. Thus, C2 of military space forces must fuse the vertical transmission of guidance, intent, direction, and status with the lateral transmission of timely and anticipatory SDA.

The C2 of military space forces applies the principles of mission command to the unique character of global space operations. As the organizing philosophy that underpins C2, *mission command* is the conduct of military operations through decentralized execution based on mission-type orders, which enables tactical-level initiative. Mission command assumes that the unit prosecuting an engagement maintains the greatest localized awareness and is best situated to rapidly identify and exploit opportunities. Mission command also assumes that in the presence of degraded or denied communications, tactical units must still be able to react on relevant timelines without prescriptive orders from higher-headquarters. In order to meet the intent of mission command, the command and control of military space forces must overcome the global and remote nature of space operations in a way that systematically provides tactical forces with the SDA required to recognize, coordinate, and exploit fleeting battlespace opportunities and prevent decision paralysis.

In accordance with the philosophical principles of mission command, the C2 of military space forces starts with an operational commander delegating responsibility and authority for a specific mission to a tactical commander through mission-type orders. Within mission-type orders, guidance and intent convey task and purpose — along with explicit constraints, restraints, and rules of engagement — and focus on the objective of the mission rather than the details of how to perform associated tasks. Mission-type orders describe *what* the mission's assigned force must do and the conditions the force must establish in order to accomplish the mission.[26] Implicit in mission-type orders is empowered mission commanders entrusted with determining *how* best to accomplish the assigned mission based on tactical judgement and battlespace awareness.

The tactical commander responsible for a mission must have the situational awareness and span of control necessary to operate in accordance with guidance and intent. Space C2, organized as a "team of teams", leverages a distributive approach to warfighter synchronization to enhance the standard centralized control structure. The C2 of military space forces requires a meshed nexus of units sharing awareness vertically and laterally, adapting to information disruptions, and harmonizing action at every level of warfare. By leveraging distributed domain awareness, military space forces are able to recognize, coordinate, and rapidly exploit transitory opportunities. Such a C2 system is robust because it enables large-scale synchronization while overcoming the fragility and cumbersomeness of centralized and hierarchical management.

CHAPTER 4 | EMPLOYMENT OF SPACE FORCES

Interoperable, reliable, and redundant communication links and information support systems are essential. Operators are trained and prepared to recognize, describe, and exploit battlespace opportunities based on operational guidance and intent. Finally, operators must be capable of communicating these opportunities vertically and laterally to other units, who they are teamed with in a force package, to execute their assigned missions. These force packages can be composed of space high value asset units, defensive units, offensive units, ISR units, and/or cyber protection units. Properly implemented, the command and control of these military space forces fosters disciplined initiative, responsiveness, tactical creativity, and interdependent action without sacrificing the centralized synchronization of capabilities that are inherently global. Such a C2 structure empowers tactical commanders with the awareness required to make decisions and assume risk even during remote operations. These characteristics allow for tactical boldness and a level of agility not historically seen within industrial-era control systems.

STEWARDS OF THE DOMAIN

Military space forces must be responsible stewards of the space domain. When designing missions, training, and performing end of life operations, military space forces should make every effort to promote responsible norms of behavior that perpetuate space as a safe and open environment in accordance with the Laws of Armed Conflict, the Outer Space Treaty, and international law, as well as U.S. Government and DoD policy. Just like all forms of warfare, the prosecution of space warfare and the potential generation of collateral damage is judged against the principles of military necessity, distinction, and proportionality. Through this approach, military space forces balance our responsibilities for operational readiness with the safety and sustainability of the space environment for use by future generations.

⬥ ⬥ ⬥

The employment of military spacepower preserves the prosperity and security the United States derives from the space domain. The grand formulation of national spacepower — complete with diplomatic, informational, military, and economic components — expands the scope and scale of space control beyond military objectives. To this end, Preserving Freedom of Action in space is an operational imperative across the conflict continuum.

Additionally, the employment of military space forces must enable the lethality of the Joint Force and provide national leadership with independent options for achieving national objectives. However, any loss of space domain freedom of action compromises the other two responsibilities. *Thus, preserving freedom of action in space is the essence of military spacepower and must be the first priority of military space forces.*

From a broader perspective, understanding the employment of military space forces answers elemental questions about the nature of military spacepower. The three cornerstone responsibilities of military spacepower — Preserve Freedom of Action, Enable Joint Force Lethality and Effectiveness, and Provide Independent Options for national leadership — answer *why* military spacepower is vital to U.S. prosperity and security. The range of spacepower core competencies — Space Security, Combat Power Projection, Space Mobility and Logistics, Information Mobility, and Space Domain Awareness — answer *how* military spacepower is employed. Taken together, the answers to these two questions shape and define the purpose and identity of military space forces, laying the foundation for a unifying culture.

War is neither a science nor a craft, but rather an incredibly complex endeavor which challenges men and women to the core of their souls. It is, to put it bluntly, not only the most physically demanding of all the professions, but also the most demanding intellectually and morally.[27]

Williamson Murray, 2011
War, Strategy, and Military Effectiveness

The stars will never be won by little minds; we must be as big as space itself.[28]

Robert Heinlein, 1956
Double Star

CHAPTER 5
MILITARY SPACE FORCES

The technology that enables orbital flight can sometimes obscure the most important component of spacepower: *our people*. Innovation propels spacepower forward; for this attribute, there can be no substitute. No amount of funding nor reorganization can compensate for a lack of innovation. While innovation can ignite the development of new technology, superior technology alone does not guarantee military dominance. The historical foundation and future determinant of U.S. spacepower is the expertise of those visionary pioneers dedicated to the many applications of orbital flight.

DEVELOPING SPACEPOWER EXPERTISE

Our greatest assets are the men and women — the space professionals — who develop, employ, and advance spacepower for the Nation. Sound doctrine and superior capabilities are of little use without personnel who have the expertise and empowerment required to wield them. It is of upmost importance that the United States Space Force prioritize the development of its people, ensuring the force is armed with the leadership, skill sets, and foresight necessary to protect and defend United States interests in any strategic or operational environment it faces.

CHAPTER 5 | MILITARY SPACE FORCES

The complexity and dynamic nature of space operations demand a range of disciplines that must work in tandem to ensure our core competencies are sustained and our cornerstone responsibilities are met. Spacepower disciplines include operations, intelligence, engineering, acquisitions, and cyber. Space operators employ their weapon systems in accordance with principles of war to gain the advantage on the battlefield. Engineers design advanced space systems and support operational planning and execution by identifying or discovering opportunities to optimize system performance. Acquisition personnel acquire and field systems on operationally-relevant timelines that meet warfighter requirements and lead-turn threats. Intelligence experts bind all disciplines by providing foundational, tactical, and operational assessments of potential adversaries and the operating environment, and by rapidly extracting pertinent information for decision-making and action. Finally, cyber professionals perform the enduring task of defending and virtually connecting all space activities to ensure space forces can access and leverage the domain across the spectrum of conflict.

Successful integration of these disciplines requires a deliberate process that cultivates a common knowledge base, incorporates all skill sets across the core competencies, and allows a range of opportunities for leadership advancement. Space professionals must develop and maintain a global perspective to provide innovative solutions to the Joint community with effects at range. These effects may not always be from United States Space Force assets and space professionals must be sufficiently agile to leverage other interagency, Allied, civil, and/ or commercial resources as required. This process begins with the recognition that personnel conducting space operations, engineering, acquisitions, intelligence, and cyber comprise the space warfighting community and must therefore master the art and science of warfare —they are the Nation's space warfighters.

SPACEPOWER MENTALITY

National spacepower requires explorers, diplomats, entrepreneurs, scientists, developers, and warfighters. Spacepower mentality permeates the broad space community, encompassing those who research, acquire, test, launch, employ, conduct space intelligence activities, command and control, generate, or sustain military space missions. Space professionals defend the Nation's interests through the control and exploitation of the space domain.

Space professionals recognize the independent impact spacepower has on National prosperity. Our global persistence postures the Joint Force to continuously assure Allies, deter aggression, coerce competitors, and defeat adversaries. We provide the enduring vigilance that protects the United States and our Allies from strategic surprise. Due to this global persistence and enduring vigilance, space professionals are perfectly postured to provide the Joint Force global, and not just regional, perspective and capabilities. As we look to the future, our orbital presence must secure the ever expanding frontier of U.S. space interests. At their most fundamental level, space professionals seek to protect our Nation's prosperity and security.

SPACE WARFIGHTERS

Military space forces — *protectors of America's space interest* — are first and foremost warfighters who protect, defend, and project United States spacepower. Our primary purpose is to secure U.S. interests through deterrence and, when necessary, the application of force. A warfighting culture is the defining difference between operating space-based information systems and employing credible military space power. As an equal part of the Joint Force, these combat-credible forces are continuously engaged in the military competition required to deter war and counter the malign actions of strategic competitors. Should deterrence fail, military space forces are prepared to fight and win our Nation's wars, in space, from space, and to space.

CHAPTER 5 | MILITARY SPACE FORCES

Military space forces must simultaneously commit themselves to two demanding professions — warfighting and the mastery of space. This duality blends art and science and forms the core purpose, identity, and culture of the military space community.

All warfighting cultures are adversary-focused. Problems of external adaptation are defined by a thinking, competent, and lethal adversary who threatens U.S. interests while problems of internal integration focus on the perpetual pursuit of combat readiness. Innovation seeks a relative advantage over that adversary. Victory and defeat — not system availability — are the most important measures of effectiveness. Military space forces fight through uncertainty in a dynamic environment by seizing the initiative through decentralized execution and the principles of mission command. Within this culture, the imperative for victory engenders a tenacious fighting spirit and the unbreakable resolve to outmaneuver and dominate an adversary.

Warfighting is a solemn endeavor. We must never let the remote aspects of space operations dilute the solemn moral dimension of warfare. Warfighters' actions carry severe consequences. Victory secures U.S. interests and prosperity while defeat jeopardizes the political ideals the United States was founded upon. As an interdependent element of the Joint Force, failure jeopardizes the safety of warfighters around the world.

MASTERY OF SPACE

Whether on land, in the air, at sea, or in cyberspace, warfighters must develop an intuitive understanding of their domain. Military spacepower is no different. Military space forces share a kindred spirit with the pioneers who have propelled humanity into space and towards the stars. The spirit of orbital flight binds us to the larger space community through common traits, skills, and abilities.

The term *space mastery* refers to a technical understanding of the physical, network, and cognitive dimensions of space operations. In addition to the physics and engineering that enable modern space systems, space mastery also includes a predictive understanding of the interests and behaviors of civil, commercial, and foreign space actors. Space mastery is developed over time through the deliberate integration of education, training, and experience.

Space mastery makes the military space community more lethal by enhancing the speed and focus of military spacepower. These traits allow military space forces to observe, orient, and decide faster than their adversaries, rapidly converging combat power on the right objective at the right time.

As an intellectual pursuit, space warfare is both science and art. Science is the systematic organization of knowledge based on empirical evidence and falsifiable hypothesizes, while art is the application of imagination, creativity, and abstraction. The science of space warfare enables us to exploit the physics of the space domain for military advantage. Science informs our understanding of movement and maneuver, connectivity, remote sensing, and the violent transfer of energy. The art of space warfare provides insight into the human elements of warfare, including leadership, operational art, uncertainty, emotion, will to fight, adaptation, and cunning.

SPACEPOWER DISCIPLINES

Seven *Spacepower Disciplines* have emerged as necessary components of military spacepower theory: orbital warfare, space electromagnetic warfare, space battle management, space access and sustainment, military intelligence, engineering/acquisition, and cyber operations. As space warfare develops and evolves, additional disciplines will certainly emerge. However, these initial disciplines are the skills the United States Space Force needs when developing its personnel to become the masters of space warfare. The combined integration of the seven spacepower disciplines arms the military space forces with the intellectual framework required to perform our core competencies.

> ### *Spacepower Disciplines*
>
> *Orbital Warfare* – Knowledge of orbital maneuver as well as offensive and defensive fires to preserve freedom of access to the domain. Skill to ensure United States and coalition space forces can continue to provide capability to the Joint Force while denying that same advantage to the adversary.
>
> *Space Electromagnetic Warfare* – Knowledge of spectrum awareness, maneuver within the spectrum, and non-kinetic fires within the spectrum to deny adversary use of vital links. Skill to manipulate physical access to communication pathways and awareness of how those pathways contribute to enemy advantage.
>
> *Space Battle Management* – Knowledge of how to orient to the space domain and skill in making decisions to preserve mission, deny adversary access, and ultimately ensure mission accomplishment. Ability to identify hostile actions and entities, conduct combat identification, target, and direct action in response to an evolving threat environment.
>
> *Space Access and Sustainment* – Knowledge of processes, support, and logistics required to maintain and prolong operations in the space domain. Ability to resource, apply, and leverage spacepower in, from, and to the space domain.

Spacepower Disciplines continued

Military Intelligence – Knowledge to conduct intelligence-led, threat-focused operations based on the insights. Ability to leverage the broader Intelligence Community to ensure military spacepower has the ISR capabilities needed to defend the space domain.

Engineering and Acquisition – Knowledge that ensures military spacepower has the best capabilities in the world to defend the space domain. Ability to form science, technology, and acquisition partnerships with other national security space organizations, commercial entities, Allies, and academia.to ensure the warfighters are properly equipped.

Cyber Operations – Knowledge to defend the global networks upon which military spacepower is vitally dependent. Ability to employ cyber security and cyber defense of critical space networks and systems. Skill to employ future offensive capabilities.

CHAPTER 5 | MILITARY SPACE FORCES

Spacepower disciplines allow military space professionals to prosecute space warfare with speed and focus. Based on Colonel John Boyd's conception of maneuver warfare, this formulation recognizes *speed* as the rapidity of action while *focus* represents the convergence of effects on an objective.[29] The principle aim of the tactical operator is to master and apply spacepower disciplines. Doing so bestows a distinct intellectual advantage, allowing military space forces to dictate the tempo of an engagement, shatter an opponent's decision process and force a rival into a reactive or paralytic mental state.

Spacepower disciplines integrate physics, technology, employment objectives, and critical interfaces into a comprehensive body of knowledge. All disciplines consist of fundamental truths, rules-of-thumb, performance criteria, and tactics, techniques, and procedures (TTPs) or TACSOPs.

Military spacepower is most effective when operations are conducted through the combined application of all seven disciplines. For example, combat power projection must fuse the principles of orbital warfare, space electromagnetic warfare, space battle management, space access and sustainment, military intelligence, and cyber operations with the capabilities provided by engineering and acquisitions into a coherent plan of action.

ART OF SPACE WARFARE — BREADTH, DEPTH, AND CONTEXT

By itself, the science of space warfare is agnostic to its application and has limited utility. It is the art of space warfare that gives science its relevance; combined, the art and science teach us how to fight and win. Warfare is not a deterministic system, and a thinking, competent, and lethal adversary easily thwarts predictable action. Prevailing against such an adversary requires us to infuse our plans at every level of warfare with leadership, operational art, ingenuity, cunning, and audacity.

The art of warfare is the application of creativity and imagination to military operations. Studying the art of warfare sharpens our intuitive understanding of chance, risk, and reward in battle. Through experience, judgment, and intuition, the art of warfare helps us recognize the pivotal factors upon which victory hinges. Once recognized and understood, our skills in the art of warfare sharpen our creative instincts, helping us devise decisive solutions. If properly developed across our ranks, these skills can unleash a creative force that overwhelms an adversary with audaciously decisive strategy, operational design, and tactics.

We study the art of warfare by reflecting on the past. It is only through historical experience — both personal and shared histories — that we can appreciate the art of warfare and refine our creative faculties. Military history allows us to learn from the mistakes of others without suffering the consequences of failure. By dissecting how others approached their dilemmas, we sharpen our own intuition and judgement. In order to refine their skills in the art of space, military space forces must study military history in breadth, depth, and context.[i]

Studying the breadth of military history builds an appreciation for the evolution of warfare, from its earliest origins to its modern manifestation. By studying this evolution, we can discern what factors change and what factors remain constant. Studying in breadth helps space professionals understand the war's enduring nature while providing insight on how warfare's character will evolve in the future. This knowledge highlights how war's universal principles extend into space. It reinforces the role spacepower plays in the larger conflict spectrum and prepares us for what the future might hold.

Military space forces must also study select engagements, battles, and campaigns in depth. The goal of studying in depth is to understand the human element of warfare. The lack of historical space warfare does not diminish the importance of studying in depth. The purpose of such an endeavor is illuminate how the human element influences the course of violent competition. Studying warfare from any domain in depth allows a warfighter to better forecast the pressures high-intensity conflict will place on their combat responsibilities.

---

[i] *This approach for developing skills in the art of warfare through the study of military history was described in 1961 in Michael Howard's essay "The Use and Abuse of Military History." Reprinted in Parameters, Journal of the U.S. Army War College, Vol. XI, No. 1.*

CHAPTER 5 | MILITARY SPACE FORCES

Studying specific engagements in detail reveals how decisions were made, and how uncertainty, friction, and chaos influenced those decisions. This knowledge highlights how luck, timing, and biases impact the course of a military engagement. While there is no substitute for the pressure of warfighting, studying in depth is the closest approximation to reliving past experiences and fully understanding the elemental causes of victory and defeat.

Finally, military space forces must study the art of warfare in its political and social *context*. As an extension of national policy, war can only be understood within the context of the political goals the belligerents aim to achieve and the social environment within which they must operate. Warfighters must work together with diplomats to secure a better peace. In order to effectively do this, military space forces must understand how civil-military relationships shape and define warfare. Using military history to understand the political and diplomatic context for warfare helps us better understand why we fight and how policy aims constrain and restrain military operations. While important to our understanding of the art of warfare, politics alone does not determine the context of warfare. This includes studying and understanding how social context influences warfare. Economy, culture, language, ethnic history, psychology, and religion all influence why and how political entities fight. Social context illuminates an adversary's will to fight, a central component of war. Understanding how interrelationships between these factors shape the character of a nation and the social underpinnings of warfare helps us better understand adversaries and strategic competitors.

LEADERSHIP

Leaders must establish and reinforce the purpose and identity of military space forces. In peacetime, these leaders must instill a unique vision, ethos, values, and esprit de corps into assigned personnel and the collective organization. During conflict, leaders set objectives, study adversary plans, and provide direction to forces. Because of this influence, leaders at all levels must relentlessly drive a distinct warfighting culture which seamlessly integrates with the Joint Force to defeat our adversaries.

Like all military leaders, leaders of military space professionals shoulder two key responsibilities: mission execution and warfighter readiness. Geographically remote or employed-in-place operations complicate the balance between these two obligations. When leaders must simultaneously balance readiness and mission execution in the same unit, fixating on one will inevitably introduce risk to the other. Leaders must use their authority to modulate these obligations based on the operating environment and mission requirements. The goal is finding an optimal balance such that excellence in one strengthens the other. In order to achieve this balance, leaders must be intimately familiar with the strengths, weaknesses, and organizational climate of the forces under their care.

WARFIGHTING READINESS

Warfighter readiness is the ability of military forces to accomplish assigned missions within acceptable risk. Leaders ensure their subordinates have the equipment, training, environmental support, and personal resiliency to fight and prevail in conflict. The core of readiness is ensuring military space forces are prepared to prevail against any adversary based on their training, resources, and equipment condition.

Readiness hinges on warfighters who are physically, mentally, and emotionally prepared to execute their mission. Resiliency — an individual's capacity to recover from hardship or difficulties — is an often-overlooked element of readiness. The demands of warfare amplify the personal and emotional stress a member experiences. Those who are physically, mentally, or emotionally overwhelmed will struggle to adapt and thrive in dynamic and stressful environments. Thus, strengthening resiliency has a direct impact on the lethality and effectiveness of a warfighting unit. When it comes to resiliency, caring for subordinates and accomplishing the mission are one and the same. Safeguarding and strengthening resiliency extends beyond the imperative for readiness and constitutes a solemn commitment every leader must embrace.

CHAPTER 5 | MILITARY SPACE FORCES

MISSION EXECUTION

Space mastery is the hallmark that distinguishes any leader during mission execution. Every military space leader must be able to make sense of complex scenarios and be capable of distilling complexity into clarity from a remote location. They must also comprehend the effect of space capabilities or the subsequent loss of those capabilities on the Joint Force. Lastly, they must look towards the aftermath of a conflict and properly consider long-term strategic or political impacts of any one engagement or action they may take. Leadership is key to creating the value system that allows these competencies to thrive across those they lead.

The fundamental warfighting team during mission execution is a crew comprised of individuals who reflect the seven spacepower disciplines. Officer and enlisted leaders of these teams are the tactical-level leaders within our force who are entrusted with critical tactical-level and operational-level missions. These leadership teams must be able to succeed in their mission by leading their team through the fog and friction of war, adversary attacks, and equipment failures. The teams they lead will not fight alone. Instead, they will enter any engagement as part of a networked and interdependent force package, working both laterally and vertically with other units. To succeed, tactical-level leaders require a combination of situational awareness, technical mastery, and critical thinking.

AGILITY — INNOVATION — BOLDNESS

Regardless of assignment, rank, or occupational specialty, the touchstones of agility, innovation, and boldness must continually guide our actions. As a lean, mission-focused, digital Service, the United States Space Force must exemplify these traits by relentlessly pursuing new ideas, taking calculated risks, and rapidly learning from failure. Leaders must also value and reward these attributes.

Empowerment is a key component of agility, innovation, and boldness. Leaders must continually develop and empower teams capable of seizing the initiative, pursuing innovation, and taking responsibility for their actions. Training, education, and mission-type orders are the foundation of empowerment; however, empowered teams are built through active practice. Empowering small teams during day-to-day operations builds the organizational instincts required for empowered teams to thrive in warfare. The transfer of authority ignites innovation and initiative by placing the onus of critical thought on the team.

A tolerance for prudent risk-taking is inseparable from the concepts of agility, innovation, and boldness. Military space forces must be skilled managers of risk, always seeking mission accomplishment at the speed of relevance while recognizing that perfection is often the enemy of good-enough. Protracted staffing processes can lengthen decision cycles and dilute the transformative potential of proposed innovations. Leaders must continually seek the proper balance between desired capabilities and fielding schedules, between rigor and efficiency, and between deliberation and action.

Periodic failure is acceptable if we are to cultivate agile teams who ceaselessly pursue innovation and the actualization of bold ideas. A defensible decision process and disciplined approach to risk management must be the standard with which failure in pursuit of empowered innovation is judged. Importantly, mistakes during training are often the best teaching tools to evolve critical thinking capability. A fault-tolerant culture learns from training mistakes without blaming the individual. Building a culture that prizes measured risk taking will further advance the frontiers of military spacepower.

CHAPTER 5 | MILITARY SPACE FORCES

Organizational culture is difficult to define and harder to measure. At a minimum, culture describes what an organization values as a collective group. Leadership plays an important role influencing culture, but a stable culture can only flourish once organizational purpose and identity are broadly understood and accepted across the group.

The purpose of military spacepower is to preserve U.S. freedom of action in space, enable Joint Force lethality and effectiveness, and provide national leadership with independent options for generating strategic effects. This purpose, in turn, shapes our identity as equals with the other warfighters responsible for military power in the air, maritime, land, and cyber domains.

Military space forces must internalize the science and art of space warfare — we must be fluent in Kepler and Clausewitz, Maxwell and Sun Tzu, Goddard and Corbett and Mahan, as well as Newton and Liddell Hart. As an inherently technical domain, military space professionals must embrace the science and art of military spacepower, developing an identity that elevates and integrates both into a seamless warfighting culture.

UNITED STATES
SPACE FORCE

ACKNOWLEDGEMENT

We would like to thank the following members of the initial writing team for the creation of this publication:

1ST EDITION SCP-A:

Casey Beard (Team Lead), Christopher Fernengel (Deputy Team Lead), Kenny Grosselin (Lead Author and Editor), Amber Dawson, Justin Dawson, Brandon Davenport, Etan Funches, Pat Gaynor, Nic Holtz, Nate Lee, Michael Mariner, Drew Miller, Genevieve Minzyk, Peter Norsky, Josh Print, Even Rogers, Eric Snyder, Anthony Surman, and Rob Yarnes.

2ND EDITION SCP-A:

Jack Anthony, Kelly Caggiano, Steven DePalmer, Brian Goodman, and Andrew Palski

EDITORS:

Francis Doiron, Jamie Green, and Rosie Suerdieck

CONTRIBUTIONS:

The following authors made significant written contributions that played an important role shaping the ideas presented in this publication:

Donald Cox & Michael Stoiko (1958)
Ashton Carter (1984, 85, 86, 87)
S. Pete Worden & John Shaw (2002)
Charles Lutes & Peter Hays (2011)

Julian Corbett (1911)
F. H. Clauser (1946)
William Durch (1984)
Walter McDougall (1985)
John Klein (2006, 19)
David Lupton (1988)
Scott Pace (2012)
Brent Ziarnick (2015)
Troy Endicott (2016)
Joan Johnson-Freese (2016)

Steven Lambakis (2001)
John Jumper (2002)
Stephen Whiting (2002)
M. V. Smith (2002)
Paul Stares (1985, 1987)
Peter Hays (1994)
Robert Newberry (1997)
Jim Oberg (1999)
Everett Dolman (2001)
Matthew Donovan (2019)

Cover Image courtesy of NASA

REFERENCES

1. Kennedy, John F. (12 September 1962). *Moon Speech.* Rice University, Houston, TX.

2. The White House. (2019). *National Security Strategy of the United States of America.* https://www.whitehouse.gov/wp-content/uploads/2017/12/NSS-Final-12-18-2017-0905.pdf.

3. McDougall, Walter A. (1985). *The Heavens and the Earth: A Political History of the Space Age.* Basic Books, NY. P. 168.

4. Ibid, p. 169.

5. Lang, Sharon Watkins. (18 August 2016). "Project Corona: America's first photo reconnaissance satellite," https://www.army.mil/article/173155/project_corona_americas_first_photo_reconnaissance_satellite

6. *Congressional Record of the United States of America, Proceedings and Debates of the 88th Congress, First Session, Volume 109* – Part 14. (1 October 1963 – 15 October 1963). Washington, D.C. P. 19156.

7. Sellers, Jerry John. (2004). *Understanding Space: An Introduction to Astronautics.* McGraw-Hill's Custom Publishing. P. 81.

    Australian Space Academy, *Satellite Orbital Lifetimes,* available at https://www.spaceacademy.net.au/watch/debris/orblife.htm.

    King-Hele, Desmond. (1987). S*atellite Orbits in an Atmosphere: Theory and Applications Springer.*

8. National Security Council. (18 August 1958). *Statement of Preliminary U.S. Policy on Outer Space, NSC 5814/1.* Washington, D.C. https://history.state.gov/historicaldocuments/frus1958-60v02/d442.

9. Joint Chiefs of Staff. (2017). J*oint Publication 1, Doctrine for the Armed Forces of the United States, 25 March 2013, Incorporating Change 1.* https://www.jcs.mil/Portals/36/Documents/Doctrine/pubs/jp1_ch1.pdf

10 Dolman, Everett Carl. (2016). *"Seeking Strategy" in Strategy: Context and Adaptation from Archidamus to Airpower,* ed. Richard J. Bailey Jr., James W. Forsyth Jr, and Mark O. Yeisley. Naval Institute Press, Annapolis, MD. P. 9.

11 Joint Chiefs of Staff. (2017). *Joint Publication 1, Doctrine for the Armed Forces of the United States, 25 March 2013, Incorporating Change 1.* https://www.jcs.mil/Portals/36/Documents/Doctrine/pubs/jp1_ch1.pdf

12 Ibid, p. ix

13 Byman, Daniel L., Waxman, Matthew, and Larson, Eric. (1999). *Air Power as a Coercive Instrument.* Project AIR FORCE, RAND Corporation. P. 1. https://www.rand.org/pubs/monograph_reports/MR1061.html

14 Joint Chiefs of Staff. (January 2020). *DoD Dictionary of Military and Associated Terms.* www.jcs.mil, p. 64. https://www.jcs.mil/Portals/36/Documents/Doctrine/pubs/dictionary.pdf

15 Joint Chiefs of Staff. (2017). *Joint Publication 1, Doctrine for the Armed Forces of the United States, 25 March 2013, Incorporating Change 1.* https://www.jcs.mil/Portals/36/Documents/Doctrine/pubs/jp1_ch1.pdf

16 Ibid, p. I-4.

17 von Clausewitz, Carl. (1984). *On War.* Trans. Ed. M. Howard and P. Paret. Princeton University Press, Princeton, NJ. P. 75.

18 Schriever, Bernard. (19 February 1957). ICBM: *A Step Towards Space Conquest.* Speech delivered at the Astronautic Symposium. San Diego, CA.

19 The White House. (2019). *National Security Strategy of the United States of America.* https://www.whitehouse.gov/wp-content/uploads/2017/12/NSS-Final-12-18-2017-0905.pdf.

20 USSPACECOM. (2020). *4 Lines of Operations: Space Warfare, Space Domain Awareness, Space Support to Operations, and Space Service Support.* Colorado Springs, CO.

21 Joint Chiefs of Staff. (10 April 2018). *Joint Publication 3-14, Space Operations.* https://www.jcs.mil/Portals/36/Documents/Doctrine/pubs/jp3_14.pdf.

22 Curtis E. LeMay Center for Doctrine Development and Education. (2018). *Annex 3-14, Counterspace Operations,* https://www.doctrine.af.mil/Portals/61/documents/Annex_3-14/3-14-D05-SPACE-Counterspace-Ops.pdf.

23 Joint Chiefs of Staff. (22 October 2018). *Joint Publication 3-0, Joint Operations, 17 January, 2017, Incorporating Change 1.* https://www.jcs.mil/Portals/36/Documents/Doctrine/pubs/jp3_0ch1.pdf?ver=2018-11-27-160457-910

24 Ibid.

25 McChrystal, Stanley. (2015). *Team of Teams: New Rules of Engagement for a Complex World.* Penguin Random House, New York, NY.

26 Ibid, II-7.

27 Williamson Murray. (2011). *War, Strategy, and Military Effectiveness.* Cambridge University Press, New York, NY.

28 Heinlein, Robert. (2013). *Double Star.* Arc Manor, Rockville, MD. P. 91.

29 Department of the Navy. (2018). *Marine Corps Doctrine Publication 1, Warfighting, 4 April 2018, 2-19,* https://www.marines.mil/News/Publications/MCPEL/Electronic-Library-Display/Article/899837/mcdp-1.

Printed in Great Britain
by Amazon